A LABORATORY COURSE IN MICROWAVE

A LABORATORY COURSE IN MICROWAVE

By

Dr. SOMI SEBASTIAN

Deptt. of Electronics & Communication Engineering
Sree Buddha College of Engineering
Pattoor, Padanilam, Alapuzha
Kerala

UNIVERSITY SCIENCE PRESS

(An Imprint of Laxmi Publications Pvt. Ltd.)

BANGALORE • CHENNAI • COCHIN • GUWAHATI • HYDERABAD
JALANDHAR • KOLKATA • LUCKNOW • MUMBAI • PATNA
RANCHI • NEW DELHI

Published by :

UNIVERSITY SCIENCE PRESS

(*An Imprint of Laxmi Publications Pvt. Ltd.*)
113, Golden House, Daryaganj,
New Delhi-110002

Phone : 011-43 53 25 00
Fax : 011-43 53 25 28

www.laxmipublications.com
info@laxmipublications.com

Price : ₹ **50.00** *Only.* *First Edition* : 2011

OFFICES

✆ **Bangalore**	080-26 75 69 30	✆ **Chennai**	044-24 34 47 26
✆ **Cochin**	0484-237 70 04, 405 13 03	✆ **Guwahati**	0361-251 36 69, 251 38 81
✆ **Hyderabad**	040-24 65 23 33	✆ **Jalandhar**	0181-222 12 72
✆ **Kolkata**	033-22 27 43 84	✆ **Lucknow**	0522-220 99 16
✆ **Mumbai**	022-24 91 54 15, 24 92 78 69	✆ **Patna**	0612-230 00 97
✆ **Ranchi**	0651-221 47 64		

UMI-9590-050-A LAB COURSE MICROWAVE-SEB **C—2277/010/10**

Typeset at : Kalyani Computer Services, New Delhi.

CONTENTS

PREFACE

A laboratory course in microwave is an easy to use laboratory manual-cum-workbook written for use in the microwave laboratory of B.E/B.Tech communication engineering courses of Indian Universities. This easy to use book contains fourteen experiments covering the important aspects of microwave sources, devices and measurements. The experiments are arranged in such a way that it will provide a systematic way of practicing the microwave techniques for the student.

Microwave test bench is introduced first and then a careful listing of experiments follows. Each experiment contains the precise statement of objectives and a list of work to be completed within each laboratory session. Discussion of the relevant theory is not presented and the teacher is entrusted with the task. With each experiment, the student is expected to go through the relevant theory from the text book as well as few references suggested by the teacher. They are expected to give the list of suggested literature and prelab questions in advance to the students. The prelab questions may be selected from the sample questions provided at the end or prepared according to the syllabus requirements. A viva shall be conducted at the start of the laboratory session of each and every experiment to verify the proficiency of the student in the theory and procedure of the particular experiment assigned to him. Detailed tabular columns and space for calculations are also provided with each experiment so that the student can use it as a workbook in the laboratory.

I hope this book will be very useful for the students as well as the large number of young teachers/instructors practicing microwaves. Comments and suggestions are always welcome.

I wish to thank the department of electronics and communication engineering at the SASTRA University, Thanjavur for providing me an excellent and stimulating academic environment that inspired this book. I must particularly record my thanks to Ms. Ramya, R., Senior Lecturer, School of Electrical & Electronics Engineering, for the useful discussions and the great interest she has shown in this work.

I am indebted to all, and I extend my sincere gratitude to all in the Sree Buddha College of Engineering, Nooranad.

—AUTHOR

Microwave Test Bench

In microwave experiments, we use a standard experimental setup often referred as microwave test bench. It is an assembly of various microwave components as shown in figure below.

Figure 1. Microwave test bench

The test bench forms a continuous wave guide structure which guides the microwave power generated by the source through various components so that one can study the wave propagation characteristics and the processing of microwaves by various components under test. A brief account of the common devices, components, and instruments were given below.

Microwave Sources

Reflex Klystron oscillator and the Gunn diode oscillator are the widely used microwave signal generators.

Reflex Klystron

Reflex Klystron is a microwave vacuum tube which utilizes velocity modulation and transit time to generate microwave oscillations. It consists of an electron gun, a cavity resonator (anode), and a repeller.

The electron gun of the reflex klystron produces a collimated electron beam which is accelerated by a DC voltage (beam voltage) and is velocity modulated by a RF field, present inside the resonator cavity, while passing through the resonator cavity. The RF field inside the cavity is setup from the RF noise induced by the bias voltage applied to the cavity. The frequency selective nature of the cavity permits the setting up of RF oscillations inside the cavity with the resonant frequency of the cavity as the frequency of oscillation of the RF field. The velocity modulated electron beam coming out of the resonator cavity moves through the drift space toward the repeller and gets deflected back to the gap by the negative potential applied to the repeller. The fast electrons travel close to the repeller and the slow ones less close to the repeller. If we control the repeller voltage in such a way that the slow and fast electrons reach simultaneously (bunching) back at the resonator, we can force the electrons

Figure 2. 723A/B and 2K25 reflex klystron (raytheon, USA)

to release the kinetic energy to the RF field inside. This process repeats every cycle of the RF field and hence results in sustained oscillations. The optimum transit time for the electron bunch to arrive at the cavity is (n+3/4) cycles after the beam initially left the cavity, where n=0,1,2,..... We can relate the transit time and the resonant frequency of the cavity by the relation

$$t = \frac{n + \frac{3}{4}}{f_n} = \frac{N_n}{f_n}$$ where N_n is referred as the mode number.

Also it should be noted that the mode number and repeller voltage are inversely proportional. The second and third modes are the strong modes and hence we set the reflex klystron to operate in these modes in the laboratory. The klystron can be mechanically tuned with a screw, which applies vertical compression to the metal envelope.

Gunn oscillator

Another microwave source used in laboratory is the Gunn diode oscillator. Gunn diode oscillations are produced when a two terminal bulk device made of GaAs or InP is subjected to high electric fields. Under these conditions the Gunn diode works as a negative resistance device and hence can act as the part of a resonant circuit to generate oscillations in LSA mode. Stable oscillators can be designed by mounting a Gunn diode in a waveguide cavity as shown.

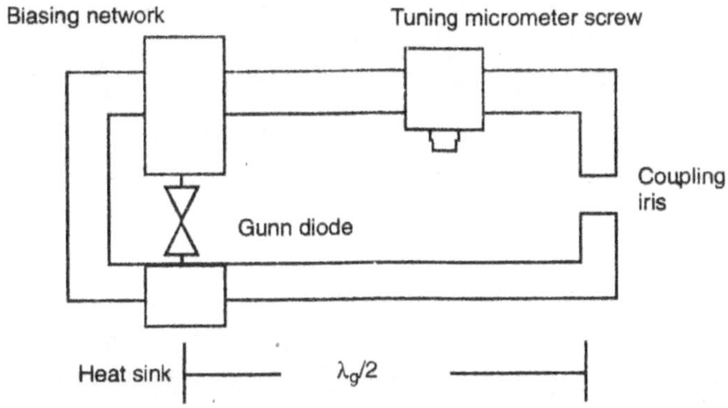

The Gunn diode oscillator is made by mounting the Gunn diode inside a waveguide cavity formed by a short circuit termination at one end and by an iris at the other end. The series equivalent circuit of a Gunn diode is a capacitance in series with a negative resistance. This capacitance may be resonated with the inductance of a shorted waveguide section of suitable length located behind the diode. The coupling to the external load may be controlled by the iris of appropriate dimensions. The distance between the diode and the iris is roughly $\frac{\lambda_g}{2}$ at the desired frequency. The Gunn diode may be mounted on a post to provide adequate heat sinking. The coupling to the waveguide is through the post. The DC bias is conveniently applied through a dumbbell filter choke combination.

Klystron Power Supply

The Reflex Klystron uses two types of regulated voltages, a high positive dc voltage called beam voltage for the accelerating grid and a high negative dc voltage for the repeller. Both the voltages can be varied. These voltages are produced separately in a regulated power supply unit. Klystron power supply unit has the provision for AM (~1 kHz) and FM (~100 Hz) modulation of beam current. Apart from this a provision for externally modulating the beam current is also generally available. A typical power supply unit is shown in figure below.

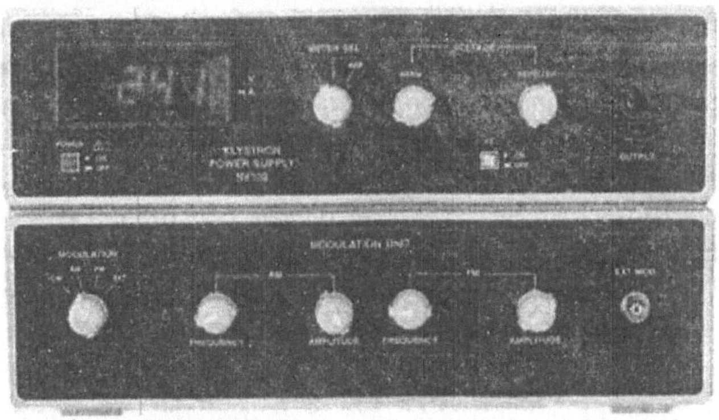

Figure 3. Klystron power supply (nvis technologies)

Initial settings of the power supply

1. Make sure that all switches are in OFF position.
2. Select the mode of operation through mode switch (CW/AM/FM).

3. Set the beam voltage control knob fully anticlockwise and repeller voltage knob fully clockwise.
4. Switch on reflex klystron power supply and cooling fan if provided.
5. Select beam voltage by using the panel meter switch and adjust appropriate beam voltage by turning the beam voltage control knob clockwise.
6. Read the beam current from the panel meter by changing the meter switch to beam current position.
7. Adjust the beam current to less than 20 mA by adjusting the beam voltage.
8. Now change the panel meter switch to repeller voltage and rotate the repeller voltage control knob slowly anticlockwise until maximum output is obtained.

Gunn Power Supply

When we use Gunn oscillator as the microwave source, the Gunn power supply provides the necessary bias voltages. It comprises a regulated DC power supply (0-10 volts) and a square wave (~1 kHz) generator. There will be a panel meter from which we can read the Gunn bias voltage and the current drawn by the diode.

Figure 4. Gunn power supply (nvis technologies)

PIN Modulator

Instead of working with continuous wave output from the microwave sources, it is better to work with audio frequencies as far as detection is considered. Thus we amplitude modulate the microwave carrier signal with ~1 kHz square wave and this permit us to use CRO, VSWR meter etc., for measurement of various parameters associated with microwave sources, devices and components. Even though a Gunn oscillator can be modulated by modulating the Gunn bias voltage, it is difficult to achieve good modulating characteristics due to various factors. Thus we use a PIN modulator operating on the continuous carrier wave generated by the Gunn oscillator. Square- wave modulation with a modulation depth of 100%, corresponds to an "ON/OFF" keying of the signal. If this modulation is to be carried out by means of a two-port device inserted between the oscillator and the remaining circuit, ideally this can be achieved by

- In the "ON" interval the two-port should be free of reflection and absorption so that the microwave signal is transmitted in its entirety.
- For the "OFF" interval the transmitted microwave signal to become zero. This can be achieved by the complete reflection of the carrier signal.

This can be carried out in a simple manner by installing electronically switchable shunt impedance into the transmission line. The magnitude of this shunt impedance should be very small in the "OFF" interval, or that the line is nearly short-circuited (reflection of the microwave signal) and very large in the "ON" interval, so that the reflection and absorption is as negligible as possible in the respective cases. The PIN modulator is a waveguide shunted with a PIN diode. A positive bias across the diode makes a short circuit in the transmission line (wave guide) and hence the waves get reflected back. A zero bias will not change the transmission characteristics and hence forward signal transmission is not obstructed.

Isolator

Isolators are used to improve the frequency stability of microwave oscillators by preventing the reflected carrier signal from reentering the source. Thus in order to ensure the smooth functioning of the microwave sources, we insert an isolator between the source and rest of the microwave bench. An isolator is a two port device which produces minimum attenuation for propagating signals in one direction and very high attenuation in the opposite direction. When the isolator is inserted between the microwave source and the load, almost all power is transmitted to the load, but the reflected signal from the load is not fedback to the source. This eliminates the undesirable variations in source output or frequency due to changing loads. Thus the isolator effectively isolates the source and the load for the reflected signal.

The amount of isolation provided by an isolator is of the order of 20–30 dB. The insertion loss is about 0.5 dB. It is usually constructed from a 3-port ferrite circulator by terminating one of its three ports with a matched load.

Frequency Meter

An absorption type frequency meter is a variable cavity resonator and is weakly coupled into the mainline through an iris on the waveguide wall and the separation between iris and cavity is approximately $\frac{\lambda_g}{2}$. The cavity is tuned to resonance by adjusting the length of the cavity using a precision piston with the drive mechanism connected to a micrometer if it is an indirect type or a drum if it is direct type. In indirect type frequency meter the frequency is found by a calibration chart which gives the relation between micrometer reading and microwave frequencies. In direct type frequency meter the frequencies are given on the drum.

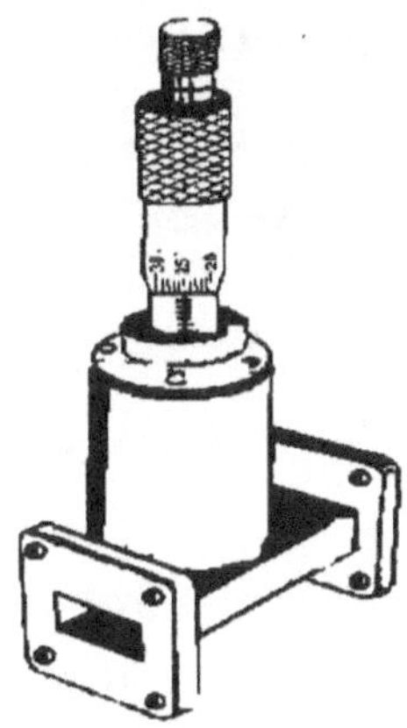

When the cavity is far from resonance it offers a short circuit at the point of connection to the waveguide and absorbs no RF power. At resonance the effective cavity impedance acts in series with the load impedance and hence absorbs RF power indicated by a dip in the output power.

Variable Attenuator

Microwave attenuators are used to control the level of signal power from the microwave source to the load to the required level. In the flap type attenuator a resistive card enters the waveguide through a slot in the broad wall, thereby intercepting and absorbing a portion of the electromagnetic wave. The amount of attenuation is varied from zero to a maximum value of about 30 dB.

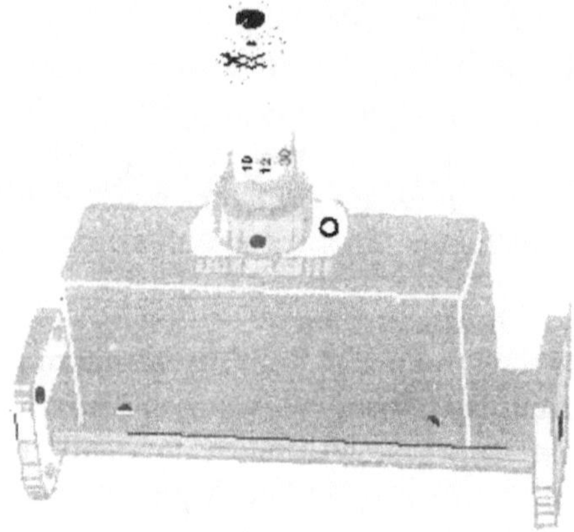

Slotted Line with Probe Carriage and Tunable Probe

A slotted line is a waveguide section (any type) with a long slot along the wall of the wave guide (in rectangular wave guides it is on the centre of broad side). The slot is parallel to flowing currents along wave guide walls, so it does not radiate from the wave guide through the slots and hence have little effect on the field inside the wave guide. The width of the slot is small enough to avoid any distortion in the original field configuration inside the waveguide. However, the slot is big enough to accommodate a coaxial E-field probe antenna mounted on a carriage which could be moved along the slot at a constant depth to achieve uniform coupling between the electric field inside the waveguide and the probe antenna in all positions. The length of the slot will be long, enough to cover a few wavelengths. A scale is fixed to the guide and a vernier scale is fixed to the probe carriage for length measurements. The probe fixed in the probe carriage generally can be moved up and down by screw mechanism to control the depth of penetration. Since the probe can be considered as an admittance shunting the line, it may function as an obstacle and disturb the field distribution inside the guide as it penetrate deep into the wave guide. To obtain accurate results with the slotted line, it is critical that the signal level be low enough so the diode is operating in the square-law region. In general, the penetrations of the sampling probe to be held at minimum. If too small, the voltage developed may not be sufficient for a measurement.

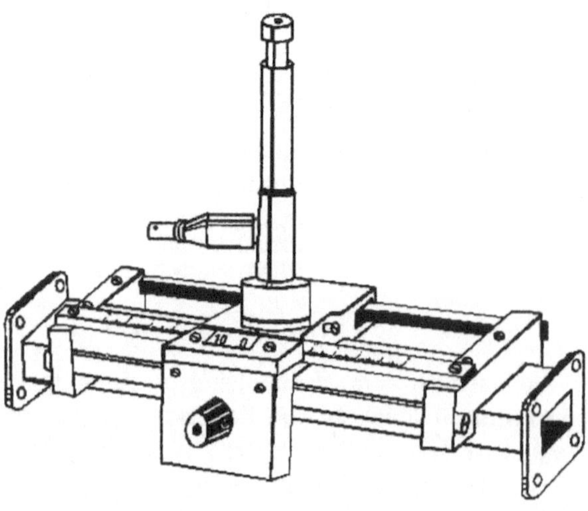

Crystal Detector

The crystal detector used for microwave detection is a microwave point contact diode rectifier. At low power levels (–20 dBm) the crystal behaves like a square law detector. The crystal is mounted inside a tunable detector mount with probe. The detector mount is a waveguide section which is flanged on one side and at the other end is affixed with a movable shorting plunger.

Square law means that the DC component of diode output is proportional to the square of the AC input voltage. Square law means that its output voltage is proportional to power on the line. So we could have a linear scale of power on the scale of the output meter. It should be noted that the output is proportional to the square of the input. This proportionality constant depends strongly on how curved is the diode's I–V curve. Schottky and germanium have very good advantage over silicon in this case. Temperature affects the I–V curve too, along with the proportionality constant. Much of the utility of our square- law detector will be in relative measurements where we need not care about the proportionality constant. All diodes will follow the square-law rule for small signals. We should choose one that is fast. In this aspect, Germanium point contact diodes and Schottky detector diodes are very good as microwave detector.

VSWR Meter

The VSWR meter is a high gain, low noise voltage amplifier tuned to a fixed frequency of 1 kHz. It is designed to work in conjunction with a slotted line or wave guide section. It is calibrated to indicate VSWR directly or the power in dB directly when it is used with a square law detector. The amplifier has a variable gain and the maximum gain possible is about 60 dB which could be altered in steps of 10 dB. There is also a coarse gain control and a fine gain control for continuous variation of gain. There are three scales for measurement of VSWR and each scale can be selected through a switch. One scale has the VSWR range from 1 to ∞ and another one from 3 to 10. For measurement of VSWR less than 1.3, there is an expanded scale which can measure VSWR from 1 to 1.3, and this ensures accuracy of measurement when VSWR is low. There is also dB scale and dB expanded scale which will be useful for measurement of high VSWR.

Preliminary settings of VSWR meter

1. Turn on the VSWR meter and wait for 3–5 minutes
2. Set input selector switch to crystal
3. Set the Gain controls (coarse and fine) to maximum position
4. Set the range selection switch to 30 dB or 40 dB position
5. Maximize the reading by adjusting any or all of the following.
   (*a*) Adjust the modulation frequency
   (*b*) Tune the diode detector
   (*c*) Adjust the probe penetration
   (*d*) Move the probe carriage along the slotted line.
6. Adjust gain controls and/or output power from the microwave source to obtain a full scale reading.

A DC ammeter or a CRO can also be used as output indicating meters in place of VSWR meter for certain measurements.

Experiment No. 1

Mode Characteristics of Reflex Klystron–I

Aim:

To study the reflex klystron oscillator as microwave source.

Task description:

1. Study the theory of microwave oscillators, recommended literature and the procedure of the experiment.
2. Write the answers for the pre-lab questions.
3. Draw the mode characteristics of the given reflex klystron.
4. Determine the mode number and transit time.

Apparatus:

1. Klystron power supply
2. Reflex Klystron with mount
3. Isolator
4. Variable Attenuator
5. Frequency meter
6. Waveguide-detector mount with detector
7. Micro ammeter, VSWR meter
8. Wave guide stands and accessories.

Procedure:

(a) Carrier wave operation

1. Setup the microwave test bench as shown in figure with micro ammeter as indicating meter.
2. Set the Klystron for CW operation with optimum beam voltage.
3. Set the repeller voltage to maximum negative value and decrease it in steps of 1V and record o/p power and frequency in table. The frequency is measured by tuning the frequency meter to have a dip in the o/p each time. The frequency meter should be detuned each time while measuring power.
4. Plot power/frequency versus repeller voltage to get mode curves.
5. By measuring the mode top voltages (repeller voltage corresponding to the mode top) of any two adjacent modes, mode number may be calculated from the equation

$$V_2/V_1 = ((n + 1) + 0.75)/ (n + 0.75)$$

6. From the mode number, transit time of each mode can be calculated as

$$t_n = (n + 0.75)/f_n$$

where f_n is the frequency corresponding to mode top of the nth mode. This can be repeated for other modes.

(b) AM wave operation

1. Setup the microwave test bench as shown in figure with VSWR meter as indicating meter.
2. Set the Klystron correctly for AM (square wave) operation with optimum beam voltage.
3. Set the frequency of modulation knob (~1 kHz) and the depth of modulation knob at mid-positions.
4. Adjust the depth of modulation and repeller voltage to obtain maximum reading in the VSWR meter and read it on the power scale.
5. Repeat steps 3 to 6 of the CW procedure.

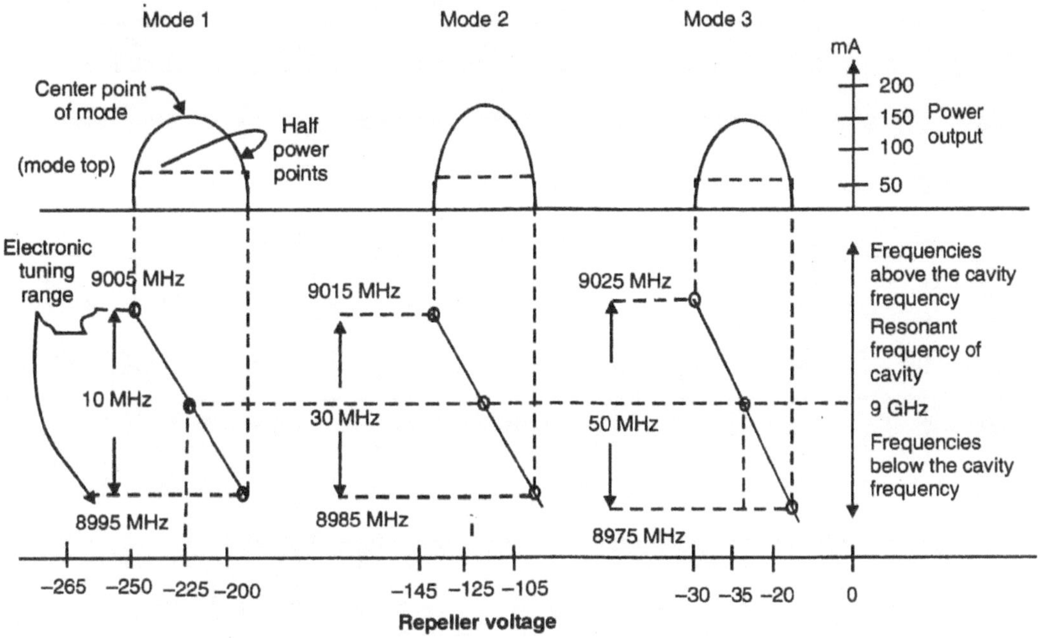

Figure 1.1. Mode characteristics of reflex Klystron

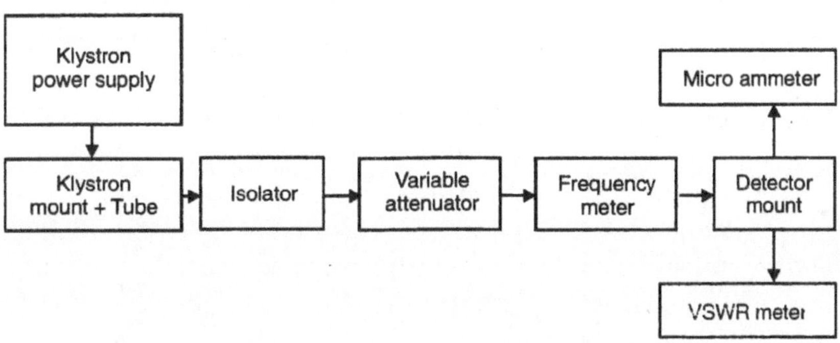

Figure 1.2. Microwave bench setup for study Klystron tube

Table1.1. Mode characteristics of reflex klystron

Sl.No.	*Repeller voltage (V)*	*Microammeter reading (μA)*	*Frequency (GHz)*
1.			
2.			
3.			
4.			
5.			
6.			
7.			
8.			
9.			
10.			
11.			
12.			
13.			
14.			
15.			
16.			
17.			
18.			
19.			
20.			
21.			
22.			
23.			
24.			

Calculations:

We have

$$V_2/V_1 = ((n + 1) + 0.75)/ (n + 0.75)$$

V_1 =.......(V) V_2 = ...(V)

n =........

From the mode number, transit time of each mode can be calculated as

$$t_n = (n + 0.75)/f_n = N_1/f_n =$$

Result:

Suggested literature for reading:

1. ....
2. ....
3. ....

Pre-lab questions:

Question No.					

Answers

Experiment No. 2
Mode Characteristics of Reflex Klystron–II

Aim:

To study the reflex klystron as a microwave source by observing the oscillogram of its mode characteristics.

Task description:

1. Study the theory of microwave oscillators, recommended literature and the procedure of the experiment.
2. Write the answers for the pre-lab questions.
3. Draw the mode characteristics of the given reflex klystron.
4. Measure the electronic band width and electronic tuning sensitivity of the given reflex klystron as explained in the procedure.

Apparatus:

1. Klystron power supply
2. Reflex Klystron with mount
3. Isolator
4. Variable Attenuator
5. Frequency meter
6. Waveguide detector mount with detector
7. CRO
8. Wave guide stands and accessories.

Procedure:

1. Setup the microwave test bench as shown in figure with CRO as indicating meter.
2. Set the Klystron correctly for FM operation with optimum beam voltage.
3. Set the frequency of modulation knob and the depth of modulation knob at mid-positions.
4. Set the CRO in XY mode.
5. Apply beam voltage and repeller voltage to get the oscillogram displayed in the CRO.
6. Adjust the magnification in such a way that only one mode is displayed in the CRO and place it symmetrically with respect to the axes (see figure 2.1).
7. Adjust the frequency meter to have dip in the mode top (figure 2.3a) and record the frequency also. Detune the frequency meter thereafter.

8. Now adjust the repeller voltage to get the oscillogram as in figure 2.3b.Change the repeller voltage in steps of 1V and record the height of the oscillogram on the Yaxis and the corresponding frequency by adjusting the dip by coinciding it with Y-axis.
9. Plot mode height/frequency versus repeller voltage to obtain the mode characteristics.
10. Now adjust the repeller voltage to get the highest mode on the oscilloscope.
11. Determine the half power points as follows: adjust the repeller voltage and the frequency meter to get the patterns of figures c,d. (make sure that the signal is symmetrically placed with respect to *x*-axis, so that the half power points falls on the x-axis as shown). Note and record the repeller voltages and the frequencies. Calculate the Electronic Bandwidth (EBW) $f_1–f_2$ and the electronic tuning sensitivity $(f_2 - f_1)/(V_2 - V_1)$

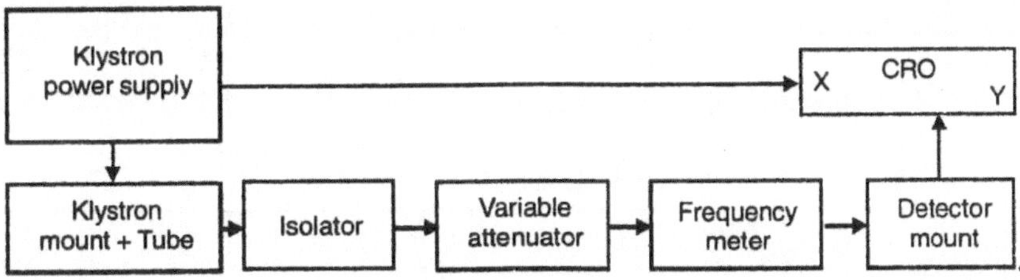

Figure 2.1. Setup for studying mode characteristics of Klystron tube using CRO

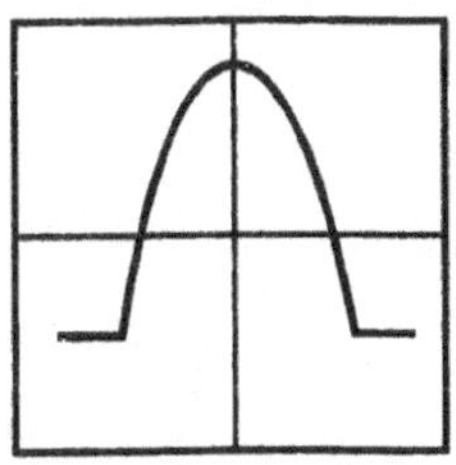

Figure 2.2. A mode pattern

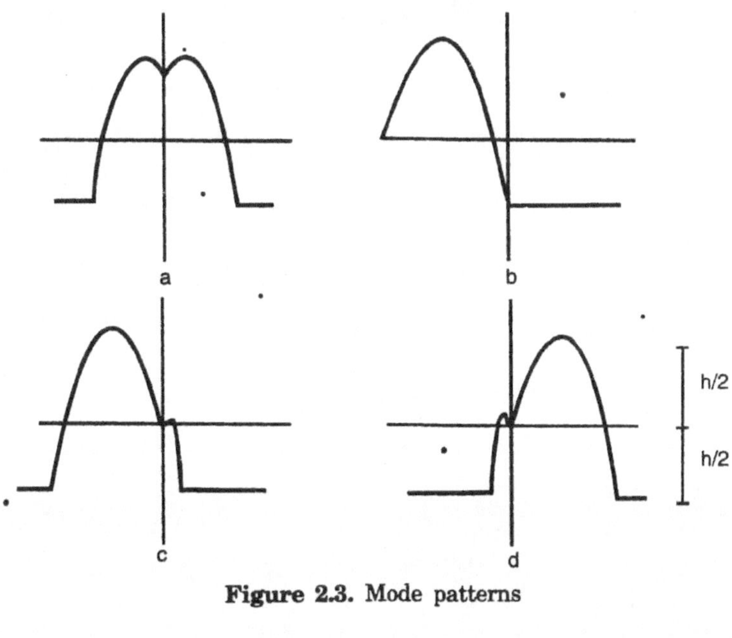

Figure 2.3. Mode patterns

Table 2.1. Mode characteristics of reflex Klystron

Sl.No.	*Repeller voltage (V)*	*Amplitude from oscillogram (V)*	*Frequency (GHz)*
1.			
2.			
3.			
4.			
5.			
6.			
7.			
8.			
9.			
10.			
11.			
12.			
13.			
14.			
15.			
16.			
17.			
18.			
19.			
20.			
21.			
22.			
23.			
24.			

Table 2.2. Electronic bandwidth and electronic tuning sensitivity

Sl. No.	*Mode*	*Repeller voltage corresponding to half amplitude*		*Frequencies corresponding to half amplitude*		*EBW*	*ETS*
		V_1	V_2	f_1	f_2		

Result:

Suggested literature for reading:

1. ....
2. ....
3. ....

Pre-lab questions:

Question No.					

Answers

Experiment No. 3
Gunn Diode Characteristics

Aim:

To study the Gunn oscillator as a microwave source.

Task description:

1. Study the theory of Gunn oscillator, recommended literature and the procedure of the experiment.
2. Write the answers for the pre-lab questions.
3. Draw the
   (*a*) I-V characteristics
   (*b*) Power/frequency versus bias characteristics
   (*c*) Power frequency characteristics
4. Measure the pushing factor.

Apparatus:

1. Gunn Oscillator
2. Gunn Oscillator power supply
3. PIN modulator
4. Isolator
5. Frequency meter
6. Attenuator
7. Detector with tunable mount
8. VSWR meter, stands, cables etc.

Procedure:

(a) I–V Characteristics

1. Setup the microwave test bench as shown in figure.
2. Keep the Gunn bias knob fully anticlockwise and set the Gunn oscillator micrometer for a suitable frequency (~9 GHz).
3. Set the variable attenuator with maximum attenuation initially.
4. Switch on the gunn power supply and VSWR meter. Adjust the VSWR meter for proper operation.
5. Change Gunn biasing in steps of 0.5 V and record corresponding currents in the table (read current in the panel meter of the Gunn power supply).
6. Draw the I–V characteristic.

(b) Power/frequency versus bias characteristics

7. Set the Gunn bias voltage above the threshold (voltage corresponding to maximum current in the I–V characteristics), set the PIN bias knob to the maximum and adjust the variable attenuator to minimum attenuation. Record power and frequency for this Gunn diode biasing.
8. Adjust the frequency of modulation to set maximum output in the VSWR meter.
9. Reduce the Gunn biasing in steps of 0.5 V and record corresponding power and frequency.
10. Draw the power versus voltage and frequency versus voltage characteristics.
11. Measure the pushing factor which is the sensitivity of the oscillator's output frequency to power supply variation. (*i.e.*, $\Delta f/\Delta V$, preferably measured around 8 volt bias)

(c) Power frequency characteristics

12. With Gunn oscillator micrometer screw in the middle of its range, set Gunn biasing for maximum power o/p and record in table this power and frequency of oscillation.
13. Move micrometer screw to one extreme till power falls to a low but readable value.
14. Move micrometer screw in the opposite direction in steps of 0.5 mm and record in table, the power and corresponding frequency till the screw reaches another extreme.
15. Plot power versus frequency characteristics and frequency versus micrometer reading curve-the calibration curve.

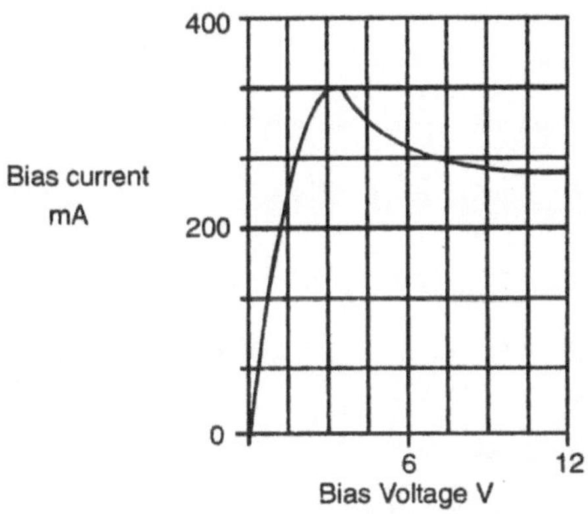

Figure 3.1. **I–V** Characteristics of the gunn diode

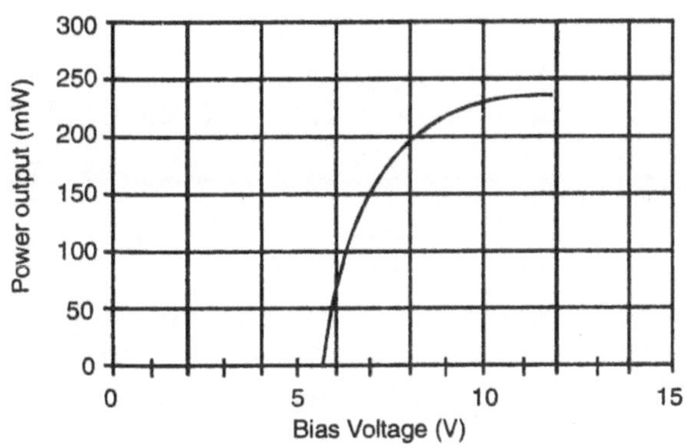

Figure 3.2. Power versus voltage characteristics

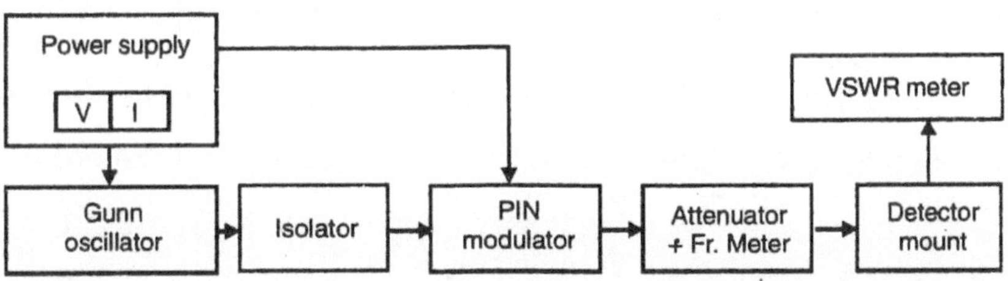

Figure 3.3. Microwave bench setup for the study of gunn diode

Micrometer Reading = Frequency =

Table 3.1. I–V Characteristics

Sl. No.	*Bias voltage (V)*	*Current (m A)*	*Threshold voltage (V)*
1.			
2.			
3.			
4.			
5.			
6.			
7.			
8.			
9.			
10.			
11.			
12.			
13.			
14.			
15.			
16.			
17.			
18.			
19.			
20.			
21.			
22.			
23.			
24.			

Micrometer reading =

Table 3.2. Power/frequency versus bias

Sl.No.	*Gunn bias (V)*	*Frequency (GHz)*	*Power (dB)*
1.			
2.			
3.			
4.			
5.			
6.			
7.			
8.			
9.			
10.			
11.			
12.			
13.			
14.			
15.			
16.			
17.			
18.			
19.			
20.			
21.			
22.			
23.			
24.			

Gunn Bias =

Table: 3.3. Power versus frequency characteristics

Sl.No.	*Micrometer reading (mm) (V)*	*Frequency (GHz)*	*Power (dB)*
1.			
2.			
3.			
4.			
5.			
6.			
7.			
8.			
9.			
10.			
11.			
12.			
13.			
14.			
15.			
16.			
17.			
18.			
19.			
20.			

Result:

Suggested literature for reading:

1. ....
2. ....
3. ....

Pre-lab questions:

Question No.					

Answers

Experiment No. 4
Slotted Line

Aim:

To study the wave guide slotted section.

Task description:

1. Study the standing waves in transmission lines, recommended literature and the procedure of the experiment.
2. Write the answers for the pre-lab questions.
3. Measure the guide wavelength and frequency of operation.
4. Measure the input VSWR of the given component.

Apparatus:

Standard microwave test bench with wave guide slotted line, blank flange and the test components.

Procedure:

(a) Wavelength and frequency of operation measurement

1. Setup the microwave test bench as shown in figure with micro ammeter as indicating meter.
2. Set the Klystron for CW operation with optimum beam voltage.
3. Tune the detector by adjusting the stub length for maximum meter deflection.
4. Measure the frequency of operation by terminating the load end with a matched termination.
5. Now terminate the load end with blank flange.
6. Move the probe along the slotted line and adjust it at minimum position (l_1), read and record in the table.
7. Move the probe to the next minimum position (l_2), read and record in the table.
8. Calculate the guide wave length as $\lambda_g = |l_1 - l_2|$.
9. Measure the length (a) of the inner broad dimension of the wave guide and compute the cut off wave length as $\lambda_c = 2a$.
10. Calculate the frequency of operation from the formula $f_0 = \frac{c}{\lambda_0} = c\sqrt{\frac{1}{\lambda_g^2} + \frac{1}{\lambda_c^2}}$ where c is the velocity of light.
11. Repeat the experiment for different frequencies of operation.

(b) VSWR measurement

1. Set the probe position for maximum reading on microammeter. That is at voltage maximum of the standing wave. Record the meter reading (Imax) in the table.
2. Move the probe along the slotted line and adjust it at minimum position, read the meter reading (I_{min}) and record in table.
3. Calculate the VSWR as $(I_{max}/I_{min})^{1/2}$.
4. Repeat the same for different loads.

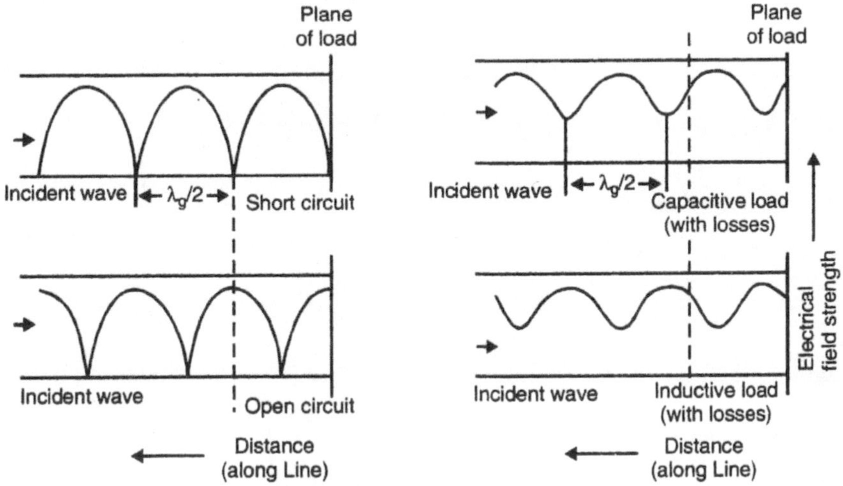

Standing wave patterns

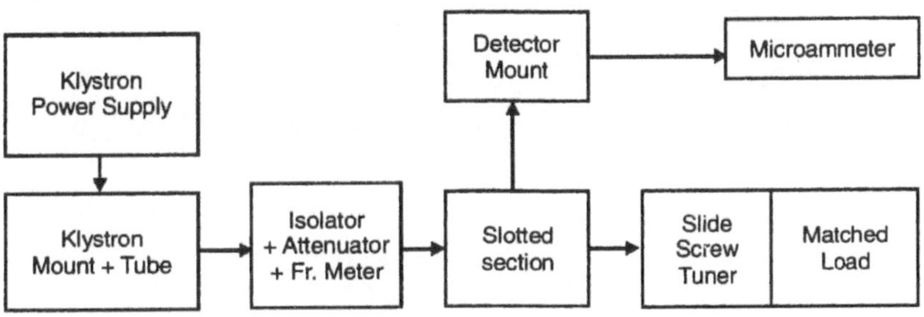

Figure 4.1. Microwave bench setup for studying VSWR

Calculation:

Table 4.1. Wavelength and frequency measurement

Frequency	l_1 *(cm)*	l_2*(cm)*	λ_g*(cm)*	λ_c *(cm)*	λ_0 *(cm)*

Table 4.2. Low VSWR measurement

Sl.No.	*Maximum reading* *[I_{max} (μ A)]*	*Minimum reading* *[I_{min} (μ A)]*	*S=*

Result:

Suggested literature for reading:

1. ....
2. ....
3. ....

Pre-lab questions

Question No.					

Answers

Experiment No. 5
VSWR Measurement

Aim:

To study the VSWR meter as microwave measurement instrument.

Task description:

1. Study the operation of VSWR meter, recommended literature and the procedure of the experiment.
2. Write the answers for the pre-lab questions.
3. Measure low VSWR.
4. Measure High VSWR.

Apparatus:

Standard microwave test bench, VSWR meter, and components under test.

Procedure:

(a) Low VSWR measurement

1. Setup the microwave test bench as shown in figure with VSWR meter as indicating meter.
2. Set the Klystron correctly for AM (square wave) operation with optimum beam voltage.
3. Set the frequency of modulation knob (~1 kHz) and the depth of modulation knob at mid-positions.
4. Adjust the depth of modulation and repeller voltage to obtain maximum reading in the VSWR meter and read it on the power scale.
5. Make sure that the probe penetration is restricted such that the VSWR value is not affected.
6. Adjust the gain of the VSWR meter amplifier to give full scale deflection to read 1.00 on the normal VSWR scale of the VSWR meter at the voltage maximum
7. Move the probe along the slotted line and adjust it at minimum position, note the VSWR meter reading.
8. If the VSWR is between 3 and 10, change the RANGE (dB) switch to next higher position and read the VSWR on second scale.
9. If the reading in the VSWR meter is less than 1.3, when the slotted line probe is in the field minimum position, select the EXPANDED scale to take readings.
10. Repeat the experiment with a different frequency as well as with different components/ loads.

(b) High VSWR measurement with VSWR meter

1. Tune the detector by adjusting the stub length for maximum meter deflection.
2. Position the probe carriage at the minimum of standing wave pattern.
3. Set the minimum meter reading as 3db in the VSWR meter.
4. Move the probe carriage by a distance less than λ/4 towards generator and record its position (l_1) in table where meter reading is 0dB.
5. Move the probe carriage towards load by a distance less than λ/2 to a position when meter reading is again 0dB. Record its position (l_2) in table.
6. Measure the guide wavelength as described.
7. Compute the SWR using $S = \lambda g/\pi \, |l_2 - l_1|$.
8. Repeat the experiment for different frequencies.

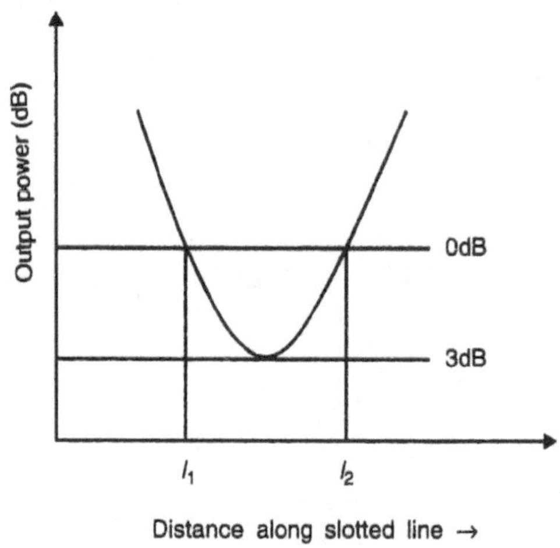

Figure 5.1. Double minima method for VSWR measurement

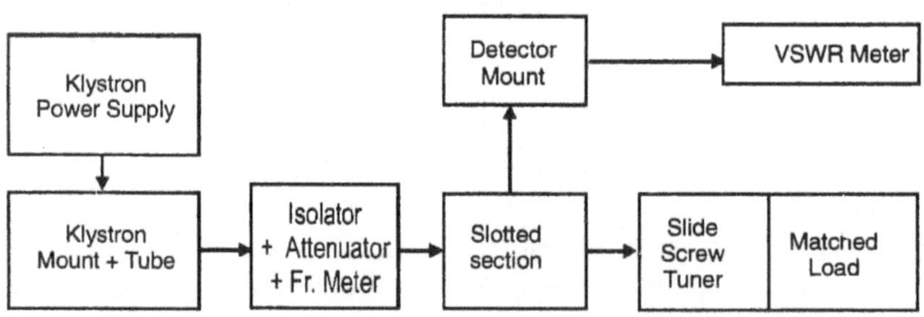

Figure 5.2. Microwave bench setup for studying high VSWR

Calculation:

Guide Wave length = Frequency of operation =

Table 5.1. High VSWR measurement

Sl.No.	*Minimum reading [I min (μA)]*	l_1*(cm)*	l_2 *(cm)*	$S = \dfrac{\lambda g}{\pi \lvert l_2 - l_1 \rvert}$

Result:

Suggested literature for reading:

1. ....
2. ....
3. ....

Pre-lab questions:

Question No.					

Answers

Experiment No. 6
Attenuator Characteristics

Aim:

To study the attenuator characteristics.

Task description:

1. Study the fixed and variable type attenuators as microwave power controllers in transmission lines, recommended literature and the procedure of the experiment.
2. Write the answers for the pre-lab questions.
3. Measure the insertion loss of the variable attenuator.
4. Measure the attenuation offered by the fixed attenuator.
5. Calibrate the given variable attenuator.

Apparatus:

Standard microwave test bench and the given fixed and variable test attenuators.

Procedure:

1. Setup the microwave test bench as shown in figure with VSWR meter as indicating meter.
2. Set the Klystron correctly for AM (square wave) operation with optimum beam voltage.

(a) Insertion loss of the variable attenuator

3. Adjust for maximum power o/p (~30 dB). Note the o/p power in VSWR meter (P_0).
4. Insert the test variable attenuator between frequency meter and detector mount without disturbing the bench.
5. Move the micrometer screw to fully released position. Note the o/p power in VSWR meter (P_1). Calculate the insertion loss as ($P_0 - P_1$).

(b) Calibration of the given variable attenuator

6. Move the micrometer screw in the opposite direction in steps of 0.5mm and note down the corresponding power (P_2) from the VSWR meter.
7. Find out the attenuation ($P_0 - P_2$) value for different micrometer readings and plot the graph (calibration curve) between micrometer reading and attenuation.

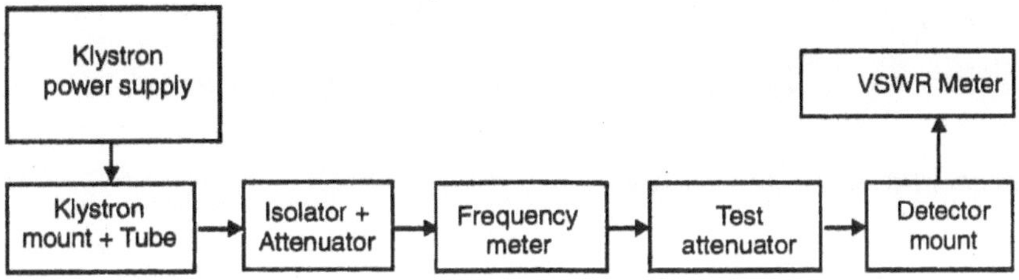

Figure 6.1. Setup for studying attenuator characteristics

Table 6.1. Attenuator calibration

Frequency of operation =

O/p power without the test attenuator, P_0 = ... (dB)

Sl.No.	*Micrometer reading (mm)*	*o/p Power (dB)* P_2	$P_0 - P_2$ *(dB)*

Calculation:

Insertion Loss of the Variable attenuator

Input power P_0 =

Output power P_1 =

Insertion loss = $P_0 - P_1$ =

Attenuation of the fixed attenuator

Input power P_0 =

Output power P_1 =

Attenuation = $P_0 - P_1$ =

Result:

Suggested literature for reading:

1. ....
2. ....
3. ....

Pre-lab questions:

Question No.					

Answers

Experiment No. 7
Isolator and Circulator

Aim:

To study the operation of isolator and circulator as microwave devices.

Task description:

1. Study the theory of microwave isolators and circulators, recommended literature and the procedure of the experiment.
2. Write the answers for the pre-lab questions.
3. Measure the insertion loss and isolation of the given test isolator and circulator.

Apparatus:

Standard microwave test bench and the given test isolator and circulator.

Procedure:

1. Setup the microwave test bench as shown in figure with VSWR meter as indicating meter.
2. Set the Klystron correctly for AM (square wave) operation with optimum beam voltage.
3. Adjust for maximum power o/p (~30 dB). Note the o/p power in VSWR meter (P_0).

(a) Insertion loss of the isolator/circulator

4. Insert the test isolator/circulator in the forward direction between frequency meter and detector mount without disturbing the bench.

   (The third port of the circulator should be terminated with a matched load.)
5. Note the o/p power in VSWR meter (P_1). Calculate the insertion loss as (P_0–P_1).
6. The same should be repeated for the other ports of the given circulator.

(b) Isolation measurement of the isolator/circulator

7. For measurement of isolation, connect the isolator/circulator in the reverse direction between frequency meter and detector mount without disturbing the bench.
8. Note the o/p power in VSWR meter (P_2). Calculate isolation as P_0–P_2.
9. The same should be repeated for the other ports of the given circulator.

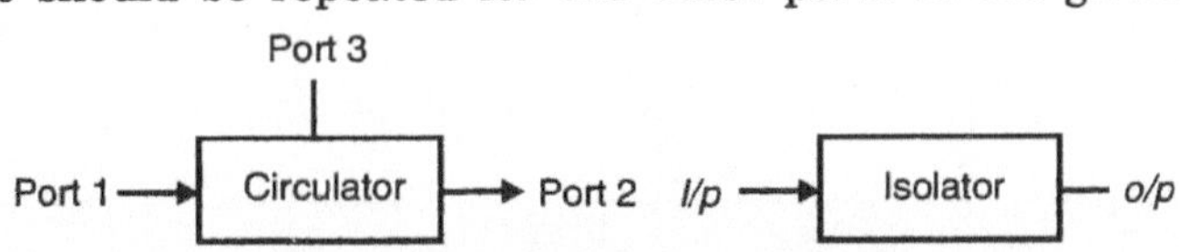

Figure 7.1. Circulator and isolator

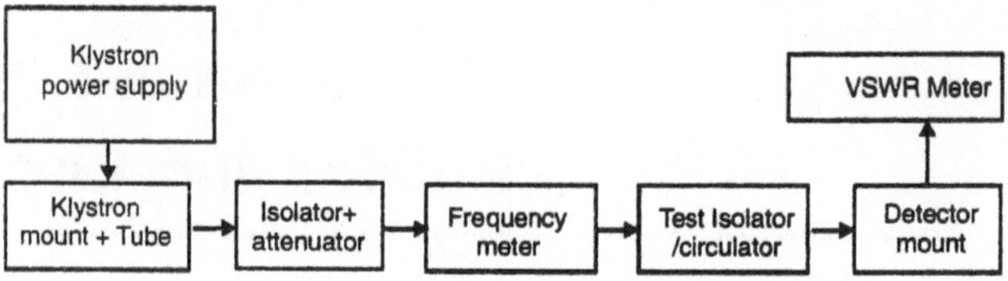

Figure 7.2. Setup for studying isolator & circulator

Table 7.1. Insertion loss

i/p port	*Power at i/p (dB)*	*o/p port*	*Power at o/p (dB)*	*Insertion loss (dB)*
Port 1		Port 2		
Port 2		Port 3		
Port 3		Port 1		

Table 7.2. Isolation

i/p port	*Power at i/p (dB)*	*o/p port*	*Power at o/p (dB)*	*Isolation (dB)*
Port 1		Port 3		
Port 2		Port 1		
Port 3		Port 2		

Result:

Suggested literature for reading:

1. ....
2. ....
3. ....

Pre-lab questions:

Question No.					

Answers

Experiment No. 8
Directional Coupler

Aim:

To study the directional coupler as microwave device.

Task description:

1. Study the multi-hole directional coupler as a sampler of microwave energy inside a waveguide, recommended literature and the procedure of the experiment.
2. Write the answers for the pre-lab questions.
3. Measure the insertion loss, coupling factor and directivity.

Apparatus:

Standard microwave test bench and the directional coupler under test and matched load.

Procedure:

1. Assemble the equipment as shown in figure with VSWR meter as indicating meter.
2. Fire the Klystron correctly for AM-MOD operation with optimum beam voltage.
3. Adjust for maximum power o/p (~30 dB). Note the o/p power in VSWR meter (P_1).

(a) Measurement of coupling factor

4. Insert the test directional coupler in the forward direction between frequency meter and detector mount without disturbing the bench and connect the detector to the auxiliary arm with the main arm o/p terminated in matched load.
5. Note the auxiliary arm o/p power in VSWR meter (P_3), *ie*., the power coupled to the auxiliary arm. Calculate the coupling factor (C) as P_1–P_3 in dB.

(b) Insertion loss

6. Now carefully interchange the detector of the auxiliary line o/p and matched load of the main line without disturbing the setup.
7. Note the main arm o/p power in VSWR meter (P_2), and calculate insertion loss (L) as P_1–P_2 in dB.

(c) Measurement of directivity

8. Restore the original arrangement, with the exception that the directional coupler under test be in the reverse direction.
9. Note the auxiliary arm o/p power in VSWR meter (P_d), It is assumed that P_i is same as before. The directivity (D) is calculated as P_3–P_d in dB.
10. Compute the isolation (I) as P_1–P_d in dB and also check I = C + D.

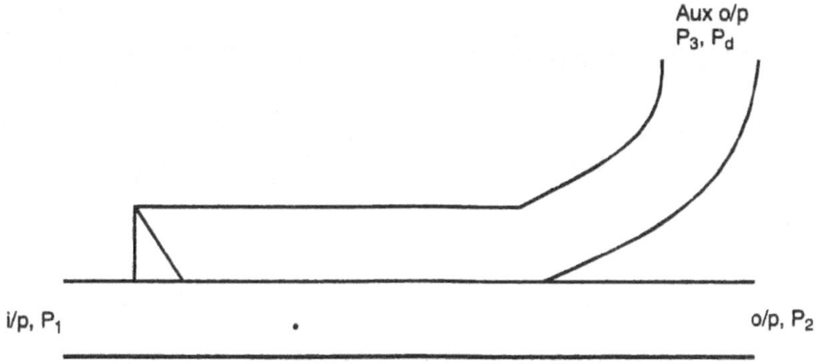

Figure 8.1. Directional coupler

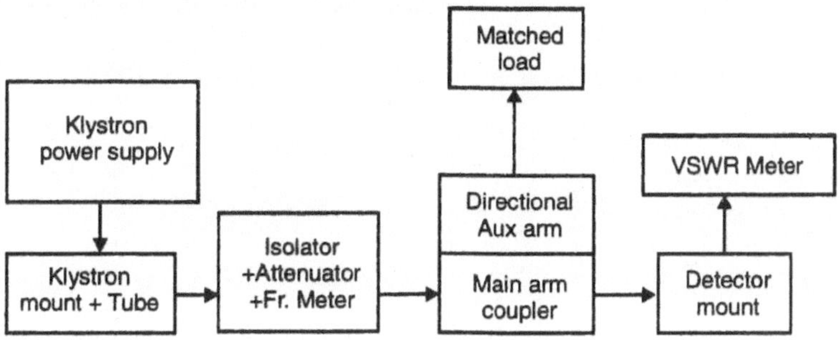

Figure 8.2. Setup for studying isolator and circulator

Calculation:

Frequency of operation =

Table. 8.1. Directional coupler

Sl.No.	P_1 (dB)	P_3 (dB)	P_2 (dB)	P_d (dB)	C (dB)	L (dB)	D (dB)	I = C + D (dB)
1.								
2.								

Result:

Suggested literature for reading:

1. ....
2. ....
3. ....

Pre-lab questions:

Question No.					

Answers

Experiment No. 9
E-plane Tee & H-plane Tee

Aim:

To study the properties of E- and H-plane T-junctions.

Task description:

1. Study the basic properties of E- and H-plane T-junctions, recommended literature and the procedure of the experiment.
2. Write the answers for the pre-lab questions.
3. Measure the isolation and, coupling coefficient between different port pairs of the given test T-junctions.

Apparatus:

Standard microwave test bench and the E- and H-plane T- junctions under test and matched load.

Procedure:

1. Assemble the equipment as shown in figure with VSWR as indicating meter.
2. Fire the Klystron correctly for AM-MOD operation with optimum beam voltage.
3. Adjust for maximum power o/p (~30 dB). Note the o/p power in VSWR meter (P_i dB).
4. Insert the test E-plane tee junction between frequency meter and detector mount without disturbing the bench and connect the detector to one of the ports and terminate the other in a matched load.
5. Determine the isolation in decibels by noting the o/p level in the VSWR meter (P_o dB). Note down the isolation as $P_i - P_o$ dB.
6. Interchange the position of the detector and matched load and determine the isolation in decibels by noting the change in the o/p level.

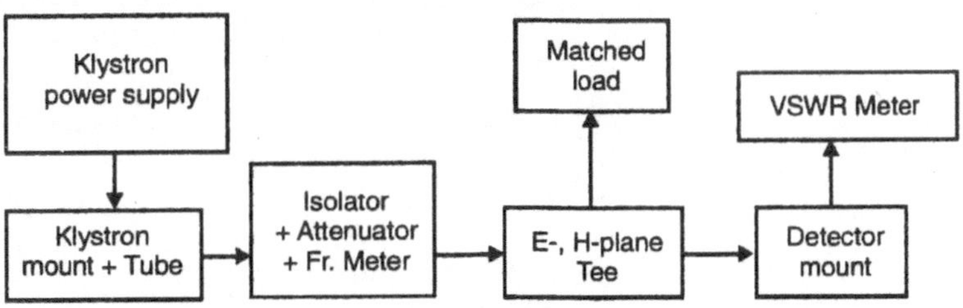

Figure 9.1. Setup for studying E-, H-plane T-junctions

7. Repeat steps 4, 5 and 6 for other orientations of the T-junction as well as for H-plane T-junction also.

8. Calculate the coupling coefficient from, $C = 10^{-\alpha/20}$ where α is the attenuation. Substitute the isolation calculated in each case as a and hence calculate coupling coefficient.

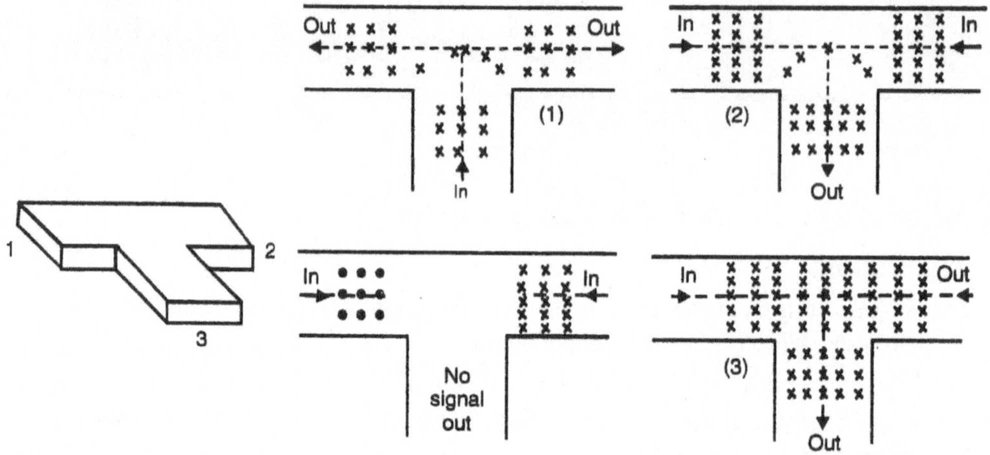

Figure 9.2 H-plane T-junction and E field inside and H-type junction for various inputs

Calculation:

Frequency of operation =

Table 9.1. Insertion loss H-plane

i/p port	*Power at i/p (dB)*	*o/p port (dB)*	*Power at o/p*	*Insertion loss (dB)*
Port 1		Port 2		I_{12} =
		Port 3		I_{13} =
Port 2		Port 1		I_{21} =
		Port 3		I_{23} =
Port 3		Port 1		I_{31} =
		Port 2		I_{32} =

$\alpha_1 = (I_{12}+I_{21})/2$ = $\qquad \alpha_2 = (I_{23}+I_{32})/2$ = $\qquad \alpha_3 = (I_{13}+I_{31})/2$

Hence the corresponding coupling factors are

$C_1 = 10^{-\alpha_1/20}$ = $\qquad C_2 = 10^{-\alpha_2/20}$ = $\qquad C_3 = 10^{-\alpha_3/20}$ =

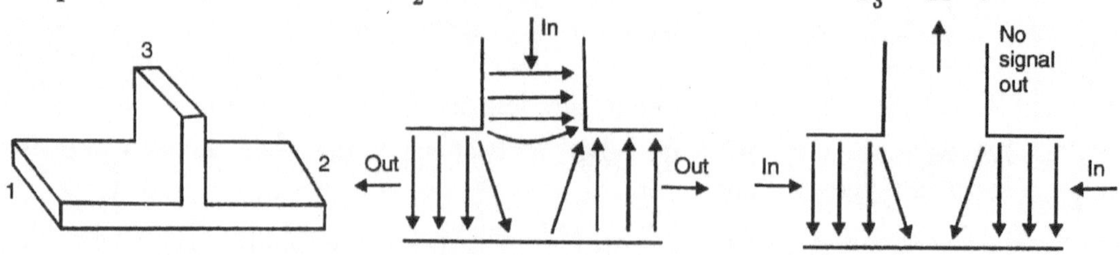

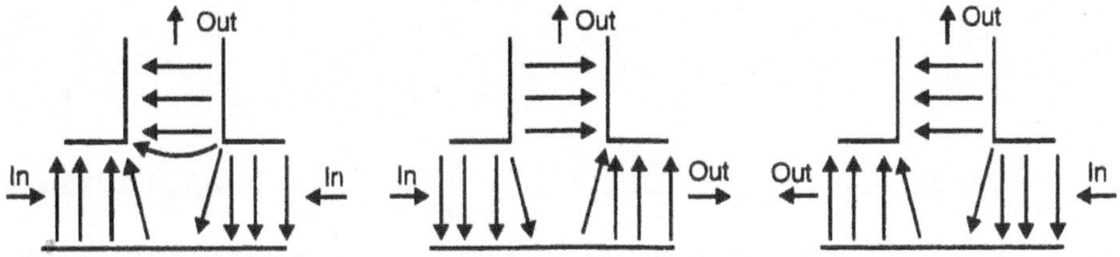

Figure 9.3. E-plane T-junction and E field inside and E-type junction for various inputs

Table 9.2. Insertion loss H-plane

i/p port	*Power at i/p (dB)*	*o/p port*	*Power at o/p (dB)*	*Insertion loss (dB)*
Port 1		Port 2		I_{12} =
		Port 3		I_{13} =
Port 2		Port 1		I_{21} =
		Port 3		I_{23} =
Port 3		Port 1		I_{31} =
		Port 2		I_{32} =

$\alpha_1 = (I_{12}+I_{21})/2 =$ $\alpha_2 = (I_{23}+I_{32})/2 =$ $\alpha_3 = (I_{13}+I_{31})/2$

Hence the corresponding coupling factors are

$C_1 = 10^{-\alpha_1/20} =$ $C_2 = 10^{-\alpha_2/20} =$ $C_3 = 10^{-\alpha_3/20} =$

Result:

Suggested literature for reading:

1. ....
2. ....
3. ....

Pre-lab questions:

Question No.					

Answers

Experiment No. 10
Magic Tee

Aim:

To study the properties of magic tee.

Task description:

1. Study the basic properties of magic-T, recommended literature and the procedure of the experiment.
2. Write the answers for the pre-lab questions.
3. Measure the isolation and, coupling coefficient between different port pairs of the given test T-junctions.

Apparatus:

Standard microwave test bench and the test magic-T and matched loads.

Procedure:

1. Assemble the equipment as shown in figure with VSWR meter as indicating meter.
2. Fire the Klystron correctly for AM-MOD operation with optimum beam voltage.
3. Adjust for maximum power o/p (~30 dB). Note the o/p power in VSWR meter (P_i).
4. Insert the test Magic tee between frequency meter and detector mount without disturbing the bench and connect the detector to one of the ports and terminate the others in a matched load.
5. Determine the isolation in decibels by noting the o/p level in the VSWR meter (P_0 dB).
6. Interchange the position of the detector and matched load and determine the isolation in decibels by noting the change in the o/p level.
7. Repeat steps 4, 5 and 6 for other orientations of the tee.
8. Calculate the coupling coefficient from $C = 10^{-\alpha/20}$, where α is the attenuation. Substitute the isolation calculated in each case as α and hence calculate coupling coefficient.

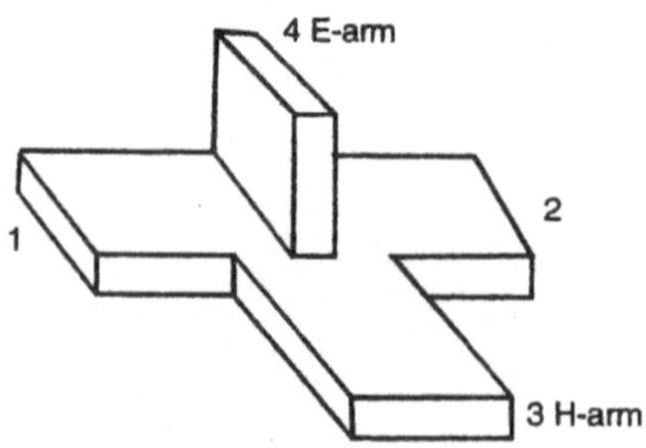

Magic Tee

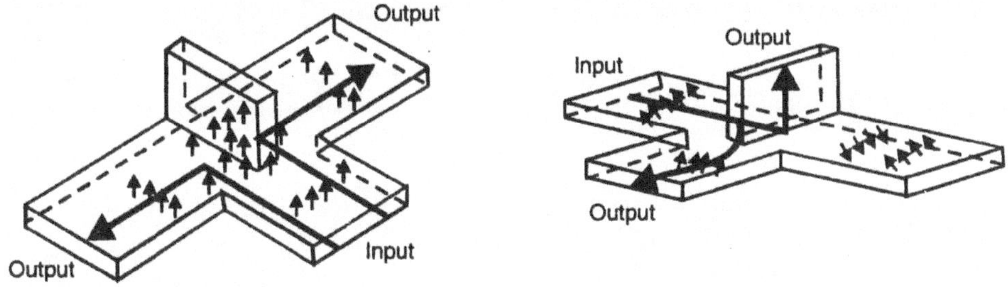

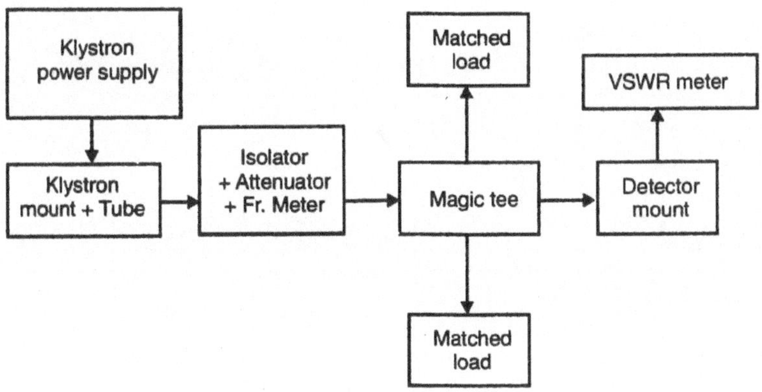

Figure 10.1. Setup for studying magic tee

Calculation:

Frequency of operation =

Table 10.1. Isolation

i/p port	*Power at i/p (P_i dB)*	*o/p port*	*Power at o/p (P_o dB)*	*Isolation P_i–P_o(dB)*
Port 1		Port 2		I_{12}
		Port 3		I_{13}
		Port 4		I_{14}
Port 2		Port 1		I_{21}
		Port 3		I_{23}
		Port 4		I_{24}
Port 3 H-arm		Port 1		I_{31}
		Port 2		I_{32}
		Port 4		I_{34}
Port 4 E-arm		Port 1		I_{41}
		Port 2		I_{42}
		Port 3		I_{43}

$\alpha_{EH} = (I_{34}+I_{43})/2 =$

Hence the corresponding coupling factor for the E and H ports is

$C_{EH} = 10^{-\alpha_{EH}/20}$

Result:

Suggested literature for reading:

1. ....
2. ....
3. ....

Pre-lab questions:

Question No.					

Answers

Experiment No. 11
Phase Shifter

Aim:

To study the phase shifter as microwave device.

Task description:

1. Study the theory of microwave phase shifters, recommended literature and the procedure of the experiment.
2. Write the answers for the pre-lab questions.
3. Calibrate the given phase shifter by plotting the calibration curve.

Apparatus:

Standard microwave test bench with slotted section, phase shifter, and movable short.

Procedure:

1. Assemble the equipment as shown in figure with VSWR meter as indicating meter and the precision movable short connected to the slotted line.
2. Fire the Klystron correctly for AM (square wave) operation with optimum beam voltage. Adjust for maximum power o/p (~30 dB).
3. Now determine the guide wavelength λg without the phase shifter inserted
4. Note a reference minimum position from the slotted line.
5. Insert the phase shifter between the slotted line and movable short with the micrometer of the phase shifter reads zero.
6. Move the short towards source end till the shift in reference minima position caused by the insertion of phase shifter, which will be indicated by VSWR meter, is nullified. It can be done by bringing back the reference minimum to its original position (reference minimum position without the phase shifter).
7. Move the micrometer screw of the phase shifter by a step of 2 mm.This will cause a shift in reference minima position which will be indicated by VSWR meter and subsequently move the micrometer of the movable short so that the reference minimum is restored to its original position indicated by the nullification of the deflection in the VSWR meter.
8. Note the micrometer reading of the movable short, which will give the equivalent phase shift corresponding to 2 mm movement of the micrometer screw of the phase shifter from its present position.
9. Repeat the experiment by moving the micrometer in steps of 2 mm and compute the corresponding phase shift using the formula:

Phase shift corresponding to $1\lambda_g$ = 360 deg
Thus phase shift corresponding to x mm shift in minimum position is;

$$\Phi = \frac{x}{\lambda_g} 360 \text{ deg}$$

10. Plot the phase shift (in degrees) versus micrometer reading of the phase shifter.

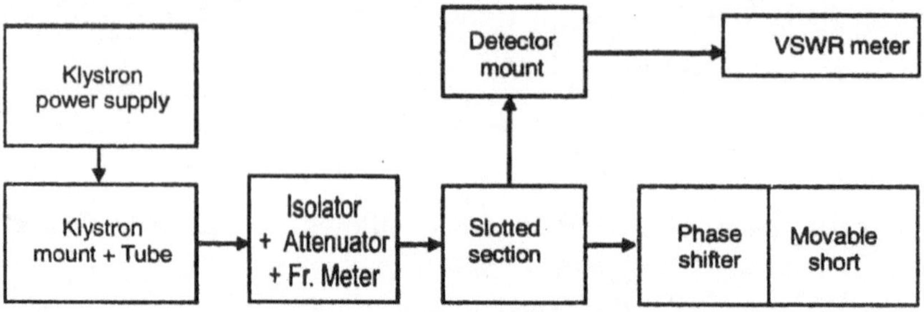

Figure 11.1. Setup for measuring phase shift

Calculation:

Guide Wave length = Frequency of operation =
Let x be the change in micrometer reading of the movable short
We know that a shift of the movable short by λ_g corresponds to 360 deg

Therefore phase shift corresponding to x, $\Phi = \frac{360}{\lambda_g} x$

Table 11.1 Phase shift measurement

Sl.No.	*Micrometer reading of phase shifter (mm)*	*Position of the movable short (mm)*	*Phase shift (Φ, deg.)*

Result:

Suggested literature for reading:

1. ....
2. ....
3. ....

Pre-lab questions:

Question No.					

Answers

Experiment No. 12
Impedance Measurement

Aim:

To familiarize the method of measurement of impedance at microwave frequencies.

Task description:

1. Study the calculation of impedance by analytical and graphical methods, recommended literature and the procedure of the experiment.
2. Write the answers for the pre-lab questions.
3. Measure the unknown load impedance analytically and by using the Smith chart.

Apparatus:

Standard microwave test bench with slotted section, unknown impedance (iris backed with a matched load-a flat plate with round hole, or a horn antenna), and blank flange.

Procedure:

1. Assemble the equipment as shown in figure with VSWR as indicating meter.
2. Fire the Klystron correctly for AM (square wave) operation with optimum beam voltage. Adjust for maximum power o/p (~30 dB).
3. Terminate the slotted section with the blank flange and accurately measure the position of the voltage minima (S_1). (see figure) .Also calculate guide wave length λg.
4. Connect the unknown impedance in place of blank flange and measure VSWR and the new positions of the minima (X_1). The VSWR can be used to calculate magnitude and the shift in position of the mimima for calculating the phase of the reflection coefficient. Note down $l_{min} = S_1\text{-}X_1$. Also determine the VSWR value S and λg.
5. Compute the unknown load impedance from the lmin, S and lg analytically as

$$Z_L = \frac{Zo(1 - js\tan\beta l_{min})}{(S - j\tan\beta l_{min})}$$ or by using smith chart.

Result:

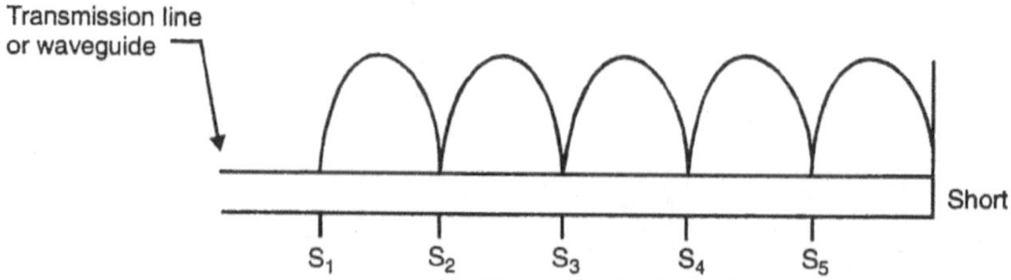

Positions of voltage minima

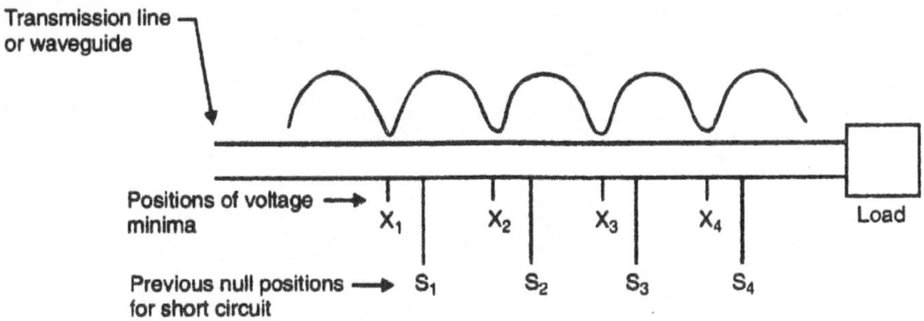

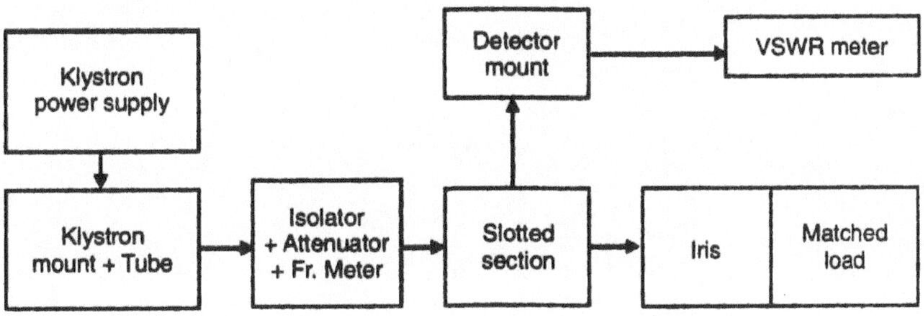

Figure 12.1. Setup for impedance measurement

Calculation:

Table 12.1. Impedance measurement

Sl.No.	S_1 *(cm)*	S_2 *(cm)*	X_1*(cm)*	λ_g *(cm)*	l_{min} *(cm)*

Guide wave length =

VSWR, S =

$$\beta = \frac{2\pi}{\lambda_g}$$

$Z_L = Z_o\ (1\text{-}jS\tan\beta\ l_{min})/(S\text{-}j\tan\lambda\ l_{min})$ =

Verification from the Smith chart

$$Z_L = Z_o \frac{(1 - jS\tan\beta l_{min})}{(S - j\tan\beta l_{min})}$$

Result:

Suggested literature for reading:

1. ....
2. ....
3. ....

Pre-lab questions:

Question No.					

Answers

Experiment No. 13
Dielectric Constant Measurement

Aim:

To familiarize the two-point method for measuring dielectric constant.

Task description:

1. Study the various dielectric measurement techniques, recommended literature and the procedure of the experiment.
2. Write the answers for the pre-lab questions.
3. Calculate the dielectric constant of the given low loss solid dielectric.
4. Calculate the dielectric constant of the given liquid dielectric.

Apparatus:

Standard microwave test bench with slotted section, samples of given dielectric, dielectric cell and blank flange.

Procedure:

Solid dielectric

1. Assemble the equipment as shown in figure with VSWR as indicating meter.
2. Terminate the slotted section with the solid dielectric cell.
3. Fire the Klystron correctly for AM (square wave) operation with optimum beam voltage. Adjust for maximum power o/p (~30 dB).
4. Adjust the movable short of the dielectric cell to the end of the cell so that it will act as a short at the end of the slotted line.
5. Now calculate guide wave length λ_g and also note down the position of reference minimum (S_1). Also note down the frequency of operation shown by the frequency meter.
6. Remove the solid dielectric cell and insert gently the dielectric sample into the dielectric cell such that one face of the rectangular shaped sample is in plane with the open face of the dielectric cell and the opposite face of the sample touches the movable short inside the cell.
7. Measure and record in table the position of the standing wave voltage minimum from the same reference minimum as chosen in step 5. Note down the position of the first minima (with dielectric) as X_1.
8. Find out the VSWR. Measure and record waveguide dimensions as well as the length of the dielectric sample.
9. Calculate the dielectric constant of the solid as.

10. The value of $\beta_\varepsilon\, l_\varepsilon$ is obtained by solving the equation $\dfrac{\tan\beta(\delta_1 + l_\varepsilon)}{\beta l_\varepsilon} = \dfrac{\tan\beta_\varepsilon\, l_\varepsilon}{\beta_\varepsilon\, l_\varepsilon}$ $= \dfrac{\tan x}{x}$, where X = β_ε, l_ε. From the tables of $\dfrac{\tan x}{x}$, select the value of X such that it gives the known value of ε_r by repeating the experiment with samples of different lengths.

$$\varepsilon_r = \frac{\frac{a}{\pi^2}\left(\frac{\beta_\varepsilon l_\varepsilon}{l_\varepsilon}\right)^2 + 1}{\left(\frac{2a}{\lambda_g}\right)^2 + 1}$$

Liquid dielectric

11. Replace the solid dielectric cell with a liquid dielectric cell.
12. Fire the Klystron correctly for AM (square wave) operation with optimum beam voltage. Adjust for maximum power o/p (~30 dB).
13. With no liquid in the cell, note down S_1, λ_g, and free space frequency and wave length.
14. Now fill the dielectric cell with the liquid and note down its depth as l_ε.
15. Measure and record in table the position of the standing wave voltage minimum from the same reference minimum as chosen in step 14. Note down the position of the first minima (with dielectric) as X_1.
16. Find out the VSWR. Measure and record waveguide dimensions and compute the value of λ_ε.
17. Calculate the dielectric constant of the liquid as

$$\varepsilon_r = \frac{\frac{a}{\pi^2}\left(\frac{\beta_\varepsilon l_\varepsilon}{l_\varepsilon}\right)^2 + 1}{\left(\frac{2a}{\lambda_g}\right)^2 + 1}$$

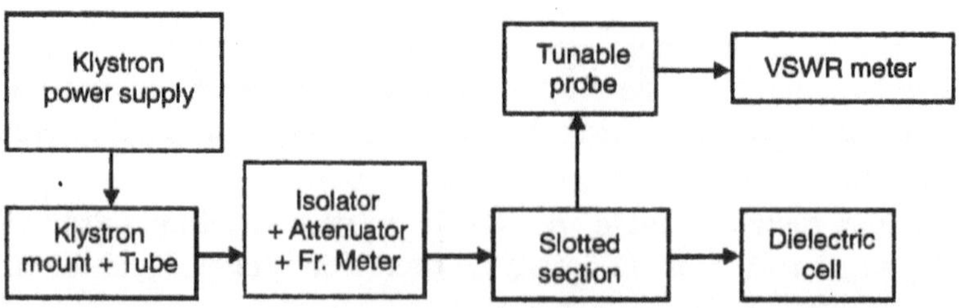

Figure 13.1. Setup for measuring the dielectric constant

Calculations

Solid dielectric

Waveguide dimension, a = b =

Cut of wavelength λ_c (= 2_a) =

Frequency of operation =

Guide wave length λ_g (without dielectric) =

Guide wave length λ_ε (with dielectric) =

Liquid dielectric

Waveguide dimension, a = b =

Cut of wavelength λ_c (=2_a) =

Frequency of operation =

Guide wave length λ_g (without dielectric) =

Guide wave length λ_ε (with dielectric) =

Table 13.1. Dielectric constant (solid)

Sl. No.	*Sample*	*Length of the dielectric l_ε(cm)*	S_1 *(cm)*	X_1 *(cm)*	$\delta_1 = S_1 - X_1$ *(cm)*	$\frac{\tan\beta(\delta 1 + l_\varepsilon)}{\beta l_\varepsilon}$	$\frac{\tan\beta_\varepsilon l_\varepsilon}{\beta_\varepsilon l_\varepsilon}$ $\frac{\tan X}{X}$	*Possible values of X and* ε_r
	I							X_1 = ε_{r1} =
								X_2 = ε_{r2} = X_3 = ε_{r3} =
	II							X_1 = ε_{r1} =
								X_2 = ε_{r2} = X_3 = ε_{r3} =
	III							X_1 = ε_{r1} =
								X_2 = ε_{r2} = X_3 = ε_{r3} =

The dielectric constant of the given solid =

Table 13.2. Dielectric constant (liquid)

Sl. No.	*Sample*	*Height of the dielectric l_ε(cm)*	S_1 *(cm)*	X_1 *(cm)*	$\delta_1 = S_1 - X_1$ *(cm)*	$\frac{\tan\beta(\delta 1 + l_\varepsilon)}{\beta l_\varepsilon}$	$\frac{\tan \beta_\varepsilon l_\varepsilon}{\beta_\varepsilon l_\varepsilon}$ $\frac{\tan X}{X}$	*Possible values of X and* ε_r
	I							X_1 = ε_{r1} =
								X_2 = ε_{r2} =
								X_3 = ε_{r3} =
	II							X_1 = ε_{r1} =
								X_2 = ε_{r2} =
								X_3 = ε_{r3} =
	III							X_1 = ε_{r1} =
								X_2 = ε_{r2} =
								X_3 = ε_{r3} =

The dielectric constant of the given replace solid with liquid =

Result:

Suggested literature for reading:

1. ....
2. ....
3. ....

Pre-lab questions:

Question No.					

Answers

Experiment No. 14
Horn Antenna

Aim:

To study the pyramidal horn antenna.

Task description:

1. Study the theory of antenna, recommended literature and the procedure of the experiment.
2. Write the answers for the pre-lab questions.
3. Calculate
   (*a*) Beam width
   (*b*) Directional gain of the antenna.

Apparatus:

Standard microwave test bench, horn antennae, waveguide twist etc.

Procedure:

1. Assemble the equipment as shown in figure with horns properly aligned and distance between antennas equal to double of $r_{min} > 2D^2/\lambda o$. E vector must be parallel to the ground.
2. Fire the Klystron correctly for CW operation with optimum beam voltage microammeter as indicating meter.
3. Adjust for maximum power o/p (~30 dB). Record the power received in the table.
4. Rotate the receiving antenna in steps of 5 deg. On both the sides read the power and enter it in the table.
5. Plot the E plane radiation pattern and determine the half power beam width θ_E. compare it with the theoretical value (θ_E. = $53\lambda o/b$).
6. Repeat the procedure after replacing the waveguide twists with a straight wave guide and compute θ_H (θ_H. = $80\lambda o/a$).
7. Compute the directional gain of the antenna using $G = 13326/\theta_E\ \theta_H$.

Gain measurement (identical horns)

1. Assemble the equipment as shown in figure with horns properly aligned and distance between antennas equal to double of $r_{min} > 2D^2/\lambda o$. E vector must be parallel to the ground.
2. Fire the Klystron correctly for AM operation with optimum beam voltage and VSWR meter as indicating meter.

3. Keep the range dB switch of VSWR meter at 50 dB position with gain control fully clockwise.
4. Obtain full-scale deflection in VSWR meter by adjusting the attenuator. This is the received power P_r.
5. Replace the transmitting horn by detector mount and change the appropriate dB position to get the deflection within the scale. Note down the power indicated by the VSWR meter. This is the transmitted power P_t.
6. Calculate the difference in dB between the P_t and P_r and convert it into a power ratio P_r/P_t.
7. The gain of the horn antenna can be calculated from the expression

$$G = \frac{4\pi r}{\lambda_0}\sqrt{\frac{P_r}{P_t}}$$

8. where r is the distance between the antennae

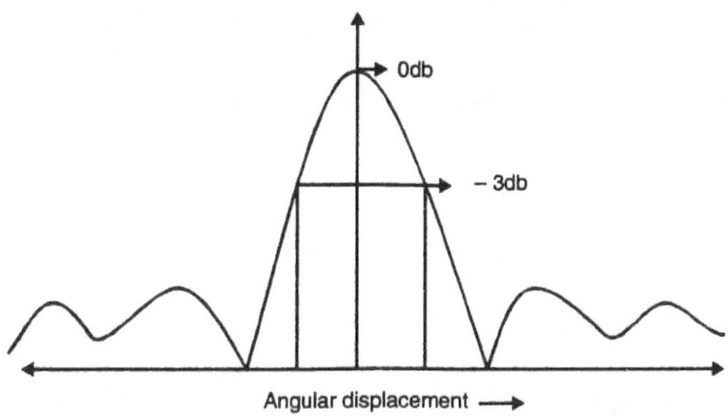

Figure 14.1. Antenna radiation pattern, rectangular plot

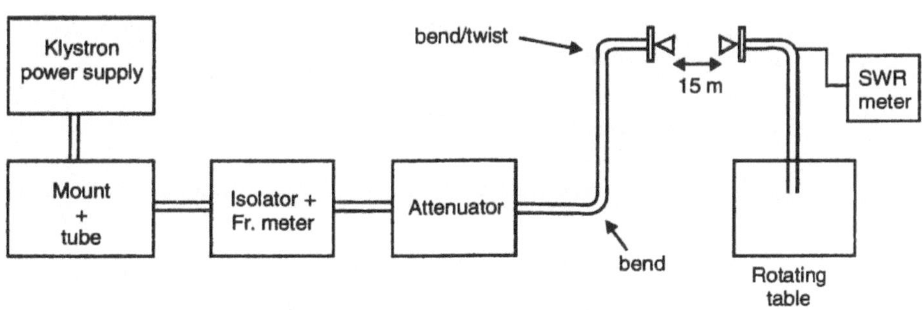

Figure 14.2. Setup for studying Horn antenna

Table 14.1. Antenna field pattern

Sl.No.	E-plane setting			H-plane setting		
	Angular setting (deg)	o/p current (μA) Left	o/p current (μA) Right	Angular setting	o/p current (μA) Left	o/p current (μA) Right
1.	5			5		
2.	10			10		
3.	15			15		
4.	20			20		
5.	25			25		
6.	30			30		
7.	35			35		
8.	40			40		
9.	45			45		
10.	50			50		
11.	55			55		
12.	60			60		
13.	65			65		
14.	70			70		
15.	75			75		
16.	80			80		
17.	85			85		
18.	90			90		

Calculations:

Waveguide dimension, a = b =

Frequency of operation =

Free space wavelength, λo =

$$\theta_E. = 53\lambda_0/b =$$

$$\theta_H. = 80\lambda_0/a =$$

$$G = 13326/\ \theta_E\ \theta_H =$$

P_t = ...dB P_r = ... dB

P_r/P_t = r=....

$$\text{Gain, } G = \frac{4\pi r}{\lambda_0}\sqrt{\frac{P_r}{P_t}} =$$

Result:

Suggested literature for reading:

1. ....
2. ....
3. ....

Pre-lab questions:

Question No.					

Answers

Microwave Radiation Hazards

Excessive exposure to microwave radiation is harmful. Although the power levels used in Microwave Laboratory are very low and should not present a health risk, it is still recommended:

- Never look into or place any part of your body against the open end of a waveguide.
- Turn off the microwave power source when assembling or disassembling components.
- Be aware of the recommended safe power limits and power levels with which you will be working.
- Develop good work ethics in the laboratory to minimize exposure to radiated fields.

Microwave radiation induces heating relatively deep inside the body to affect sensitive organs. Health risks increase according to the power density and the duration of the exposure. The eye is the most sensitive organ and cataract may develop due to exposure to microwaves. The sources used in the Microwave Laboratory, such as klystrons and Gunn diodes, deliver power outputs in the 10–15 mW range. In most cases, there is little danger of being exposed to radiation at these power levels because experiments use coaxial lines or waveguide, which provide a high degree of shielding. At the same time the power densities at the open ends of the waveguides may be high enough to cross the safety limits.

Pre-Lab/viva Questions

1. What are the limitations of vacuum tubes when it is used at high frequencies?
2. What is electron bunching?
3. What is repeller protection and why is it required?
4. Why transit time becomes a limiting factor at high frequencies?
5. Why a microwave is called "Microwave"?
6. What is the reason for the discrete modes of operation of reflex klystron?
7. Distinguish between AM wave operation and CW wave operation of a reflex Klystron.
8. Why a VSWR meter is not the preferred o/p device for CW operation?
9. Distinguish between the mechanical tuning and electronic tuning of a reflex klystron.
10. Comment on the efficiency of a reflex klystron.
11. Enlist the advantages of microwave frequencies over lower frequency waves.
12. Decode 2K25 and 1N23B.
13. Why the modulation frequency is set equal to 1 kHz?
14. What is significance of electronic tuning sensitivity?
15. What is the maximum frequency deviation in the FM employed in your system?
16. What is Velocity modulation?

17. Discuss the problem of transit time effect in tubes at microwave frequencies.
18. List two advantages of using Microwave frequency over low frequency?
19. Explain the principle of working of probes and loops.
20. In what type of application reflex klystron is preferred and why?
21. What is a square law detector? Which diodes exhibit this law?
22. What is Gunn effect?
23. What are the modes of oscillations in Gunn diode?
24. What is a Transfer Electron Devices?
25. How depletion layer is formed in PIN diode at zero bias?
26. What are the basic requirements for two-valley theory of Gunn diode?
27. Explain the V–I characteristics of a Gunn diode.
28. Explain the use of PIN diode as a modulator.
29. Which microwave diode is suitable at low frequencies and why?
30. Why silicon junction diodes are not used as microwave detectors?
31. What is the dominant mode and cut off frequency of WR90 waveguide?
32. Why TEM waves cannot be propagated in wave-guides?
33. Distinguish between probes and loops.
34. What is VSWR and reflection Coefficient? Give their minimum and maximum values.
35. What is matched load?
36. Why voltage minima are preferred over voltage maxima for a measurement?
37. Why we are modulating the RF wave with 1kHz square wave?
38. How will you make sure that the probe penetration is appropriate?
39. What do you understand by WR90, RG52 and WG16?
40. What are the inner and outer dimensions of a WR90 wave guide?
41. Give the relationship between voltage reflection coefficient and VSWR.
42. What are the values of reflection coefficient and VSWR in the following cases?
    (*a*) Matched termination (*b*) Short circuit termination
43. How will you make sure that the probe penetration would not affect the VSWR?
44. Suggest a better method to determine the voltage minimum position.
45. Differentiate between voltage reflection coefficient and current reflection coefficient
46. Define standing wave ratio.
47. What are the values of reflection coefficient and SWR of a transmission line terminated in an open circuit?
48. Give the reasoning for amplitude modulation of the RF wave with a 1kHz square wave
49. What is the importance of VSWR measurements in transmission line studies?
50. What are the possible errors in a VSWR measurement?
51. What are attenuators? Mention few applications.
52. Explain the working of an attenuator.

53. Explain the working of PIN diode attenuator.
54. What is the need for calibration?
55. Differentiate between an attenuator and an isolator.
56. Explain the working of a ferrite attenuator.
57. Explain the substitution method for attenuation measurement.
58. Explain the working of a rotary attenuator.
59. What are the properties that make the Ferrite devices useful in microwave applications?
60. What is Faraday rotation?
61. Why ferrite devices are unidirectional?
62. Explain how circulators are designed with ferrites.
63. What are the important applications of circulators?
64. Define Faraday's rotation and also identify any device based on faraday's rotation.
65. What is the primary purpose of a directional coupler?
66. How far apart are the two holes in a simple directional coupler?
67. What is the purpose of the absorbent material in a directional coupler?
68. What happens to reflected energy that enters a directional coupler that is designed to sample incident energy?
69. Define Directivity in case of a directional coupler.
70. What do you understand by a 3 dB directional coupler?
71. Define the Coupling factor and Isolation in the case of directional coupler.
72. What is an E-Plane tee?
73. Explain the working of H-plane tee.
74. Distinguish between ~~E-plane~~ and H-plane tee.
75. What are the two basic types of T junctions?
76. Why is the E-type T junction so named?
77. Write the S-matrix for series tee and shunt tee.
78. Draw the electric field inside the series tee when the input port is the series port?
79. Why there is no phase difference between the collinear arms in a shunt tee?
80. The magic-T is composed of what two basic types of T junctions?
81. What are the primary disadvantages of the magic-T?
82. Hybrid rings are used primarily for what purpose?
83. What is a rat-race?
84. What is a duplexer?
85. What is magic Tee and why E-H plane is named as magic tee?
86. What is the advantage of rat-race over magic-T?
87. What do you understand by IEEE Std. C95.1 and C95.3?
88. What is the Maximum Permissible Exposure at 10 GHz?
89. Differentiate between the controlled environment and the uncontrolled environment defined by the IEEE standard for safety levels with respect to human exposure to radio frequency electromagnetic fields.

Index